Venus

by Steven L. Kipp

Content Consultants:
Rod Nerdahl
Program Director, Minneapolis Planetarium

Diane Kane
Space Center Houston

Bridgestone Books

an imprint of Capstone Press

Bridgestone Books are published by Capstone Press
818 North Willow Street, Mankato, Minnesota 56001
http://www.capstone-press.com

Library of Congress Cataloging-in-Publication Data
Kipp, Steven L.
    Venus/by Steven L. Kipp.
    p. cm.--(The galaxy)
    Summary: Discusses the orbit, atmosphere, surface features,
exploration, and other aspects of the planet Venus.
    ISBN 1-56065-609-3
    1. Venus (Planet)--Juvenile literature. [1. Venus (Planet)]
I. Title.   II. Series: Kipp, Steven L.   Galaxy.

QB621.K57   1998
523.42--dc21

                                            97-6923
                                            CIP
                                            AC

Photo credits
Steven L. Kipp, 6
NASA, cover, 8, 10, 12, 16, 18, 20

Table of Contents

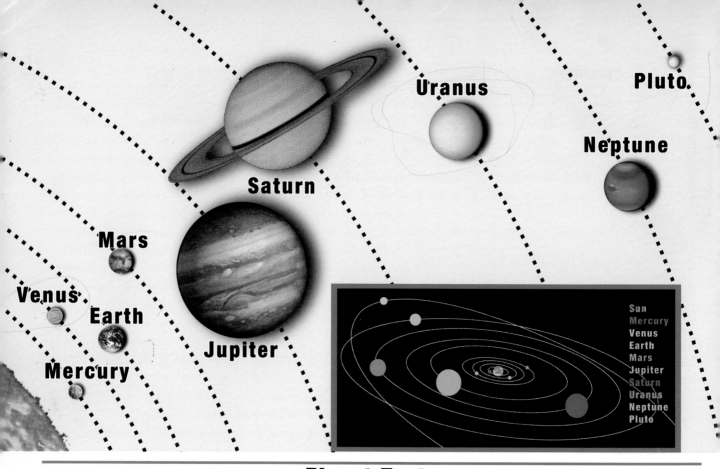

Saturn

Uranus

Pluto

Neptune

Mars

Jupiter

Venus

Earth

Mercury

Sun
Mercury
Venus
Earth
Mars
Jupiter
Saturn
Uranus
Neptune
Pluto

Planet Facts

Venus

Diameter–7,521 miles (12,401 kilometers)

Distance from Sun–67 million miles (107 million kilometers)

Moons–Zero

Revolution Period–225 days

Rotation Period–243 days

Earth

Diameter–7,927 miles (12,756 kilometers)

Distance from Sun–93 million miles (150 million kilometers)

Moons–One

Revolution Period–365 days

Rotation Period–23 hours and 56 minutes

Venus and the Solar System

Venus is part of the solar system. The solar system includes the Sun, planets, and objects traveling with them. The solar system is always moving.

The Sun is the center of the solar system. Everything in the solar system circles around the Sun. The Sun is a star. A star is a ball of very hot gases. Stars like the Sun give off heat and light.

There are nine known planets in the solar system. Planets are the nine heavenly bodies that circle the Sun. Venus is one of them.

Venus is the second planet from the Sun. It is about 67 million miles (107 million kilometers) away from the Sun. Venus circles the Sun at the speed of 22 miles (35 kilometers) a second. Venus rotates as it travels, too. This means it spins.

Planets receive heat from the Sun. Venus is closer to the sun than Earth. It becomes very hot there. So far, scientists have not found any living things on Venus.

Venus

Planet Venus

Venus is one of the brightest objects in the sky. Only the Sun and Earth's Moon are brighter. People can see Venus without a telescope. A telescope makes faraway things seem larger and closer.

Early Italian people liked this bright planet. They named the planet Venus after their goddess of love. Back then, people did not understand Venus. They thought it was a star like the Sun. Later, scientists learned that Venus is a planet.

Sometimes people see Venus in the western sky after sunset. Then they call Venus the evening star. Sometimes people see Venus in the eastern sky before sunrise. Then they call Venus the morning star.

Venus and Earth look very different from each other. Earth looks blue from space. Venus looks whitish-yellow from space.

Venus is one of the brightest objects in the sky.

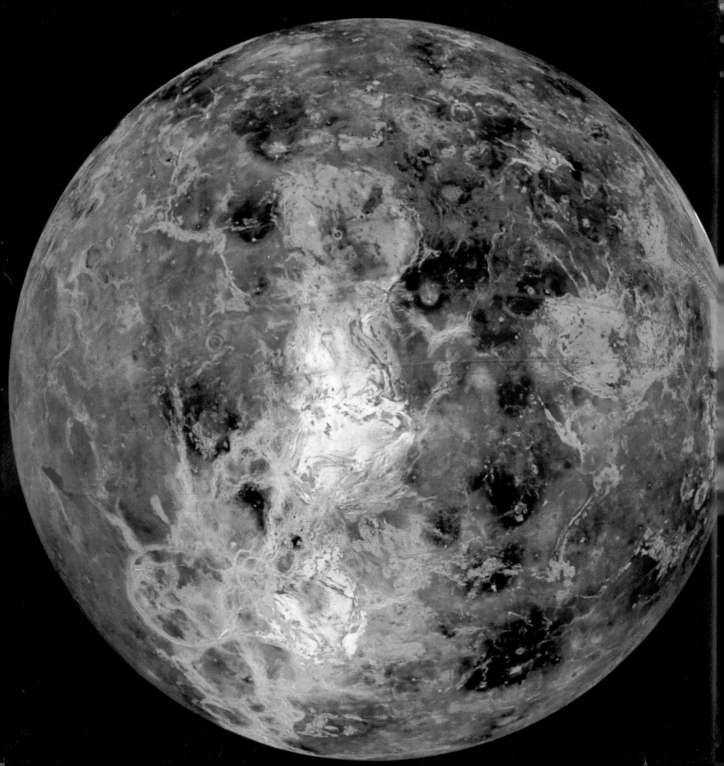

Twin Planet

Sometimes people call Venus the twin planet. This is because scientists once believed Venus and Earth were alike.

Scientists thought Venus might have oceans like Earth. They thought Venus had an atmosphere like Earth's, too. Atmosphere is the mix of gases that surrounds some planets. Scientists learned these things were not true.

Venus does have an atmosphere. But its atmosphere has different gases than Earth's atmosphere. Venus has no oceans. In fact, it has no water at all.

Venus is like Earth in one way. The planets are about the same size. Earth is 7,927 miles (12,756 kilometers) wide. Venus is 7,521 miles (12,104 kilometers) wide. Venus is only a little smaller than Earth.

Venus is 7,521 miles (12,104 kilometers) wide.

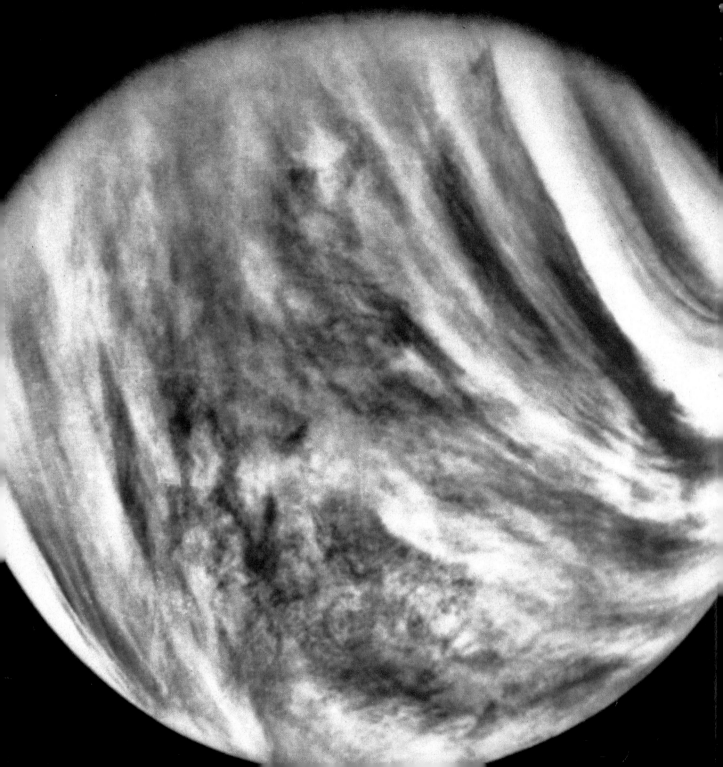

Atmosphere

The atmosphere of Venus is made of gases and dust. It is mostly made of a gas called carbon dioxide. People breathe out carbon dioxide and breathe in oxygen. Atmosphere on Venus has many other gases. But it has very little oxygen. People could not breathe on Venus.

Thick clouds form part of the atmosphere of Venus. On Earth, water helps form clouds. On Venus, sulfuric acid forms clouds. Sulfuric acid can burn skin. On Venus, it rains sulfuric acid instead of water.

Venus' atmosphere is very thick. It is like a weight pressing things down. The atmosphere's weight pressing on things is 90 times heavier than on Earth. This makes objects on Venus heavier. A person who weighs 100 pounds (40 and one-half kilograms) on Earth would weigh 9,000 pounds (4,050 kilograms) on Venus. A person could not live under this weight.

Thick clouds form part of the atmosphere of Venus.

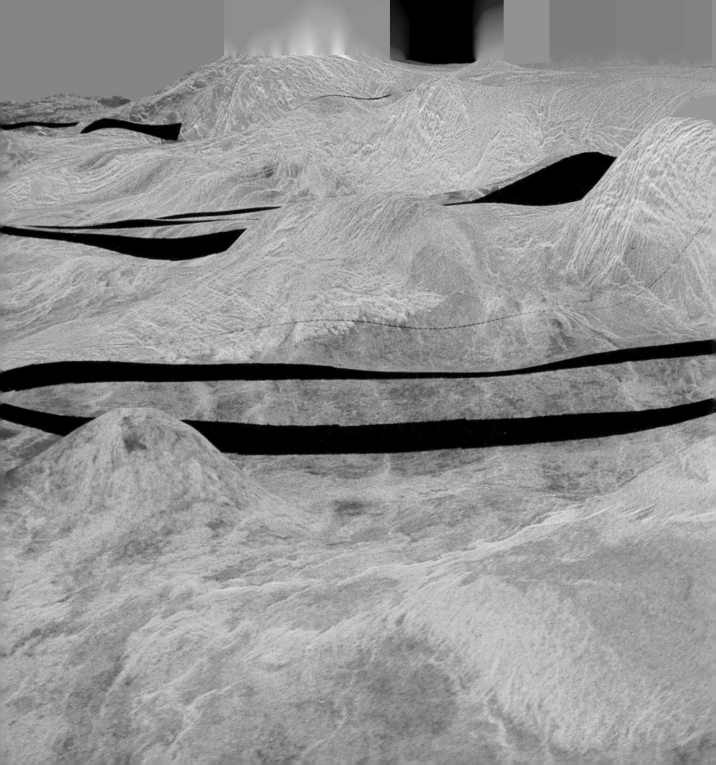

Greenhouse Effect

Scientists once thought Venus had the same range of temperatures as Earth. Temperature is the amount of hot or cold in something. But the temperatures on Venus are very different than Earth's temperatures.

Scientists discovered that the average temperature on Venus was about 900 degrees Fahrenheit (482 degrees Celsius). This is 400 degrees hotter than most ovens can reach. Venus is the hottest of all the planets. Venus has only land. Liquid cannot exist on a planet with such hot temperatures.

Venus' thick, carbon dioxide atmosphere causes its high temperature. It acts like a blanket. It only lets a little heat escape. This is called the greenhouse effect. It happens when heat from the Sun reaches Venus. Then the planet's atmosphere holds in most of the heat.

Liquid could not exist on a planet with such hot temperatures.

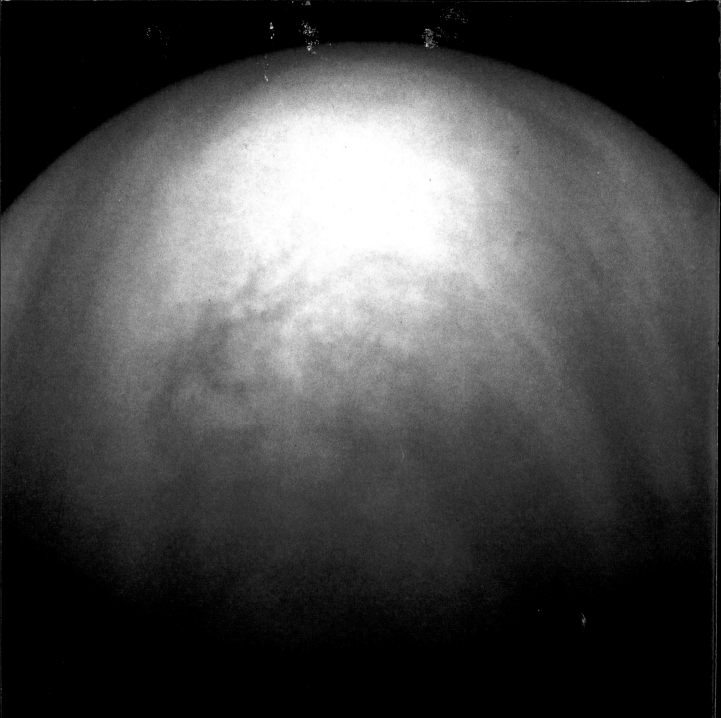

Rotation and Revolution

Venus spins as it moves through space. One complete spin is called a rotation. Rotation time makes up a planet's day. Venus spins very slowly. Earth spins around once about every 24 hours. It takes Venus 243 Earth days to complete one spin. This is the slowest rotation time of any planet. One day on Venus is about eight months long on Earth.

The rotation of Venus is different in another way. It rotates from west to east. The Sun rises in the west on Venus. It sets in the east. On Earth, the opposite is true.

Venus circles the Sun like Earth does, too. It does this in a path called an orbit. One complete orbit around the Sun is called a revolution. Revolution time makes up a planet's year. It takes Venus 225 Earth days to circle the Sun. It takes Earth 365 days to circle the Sun. A year on Venus is shorter than its day.

Venus has the slowest rotation time of any planet.

Phases and Transits

Astronomers are people who study stars, planets, and space. Galileo Galilei was a famous astronomer who lived during the 1600s. He was the first scientist to look at Venus with a telescope.

Galileo discovered that Venus seemed to change shape like Earth's Moon. But Venus does not really change shape. The Sun lights different parts of it. From Earth, people can see only the part of Venus that the Sun lights. The part that people can see is called a phase. Earth's moon shows phases, too.

Sometimes Venus passes between the Sun and Earth. This is called a transit. Transits of Venus are very uncommon. The next transits will be on June 8, 2004, and June 6, 2012.

Venus shows phases as the Sun lights different parts of it.

Photographing Venus

Russian Venera spacecrafts landed on Venus in the 1970s. Venera is the Russian name for Venus. A spacecraft is a craft built to travel in space. Spacecrafts help scientists learn. The Venera spacecrafts were the first spacecrafts to land on Venus. They took pictures of Venus for scientists.

The pictures showed that flat rocks with sharp edges cover Venus. Venus is also covered with craters. Craters are large holes in the ground. They were made when meteorites crashed into Venus. Meteorites are large rocks from space that hit a planet's surface.

The sunlight on Venus looks yellow-orange. This makes the surface look orange, too. But rocks are not really orange on Venus. The sunlight that travels through the clouds seems orange. The sky on Venus always looks orange. It is the color of a sunrise or sunset on Earth.

Craters cover parts of Venus.

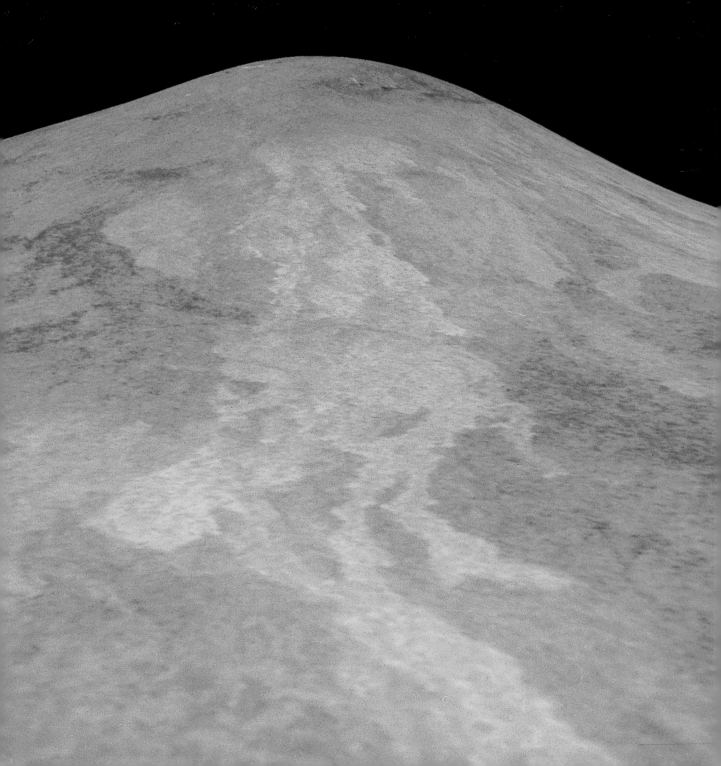

On Venus

There are many large volcanoes on Venus. Volcanoes are mountains that form over rips in a planet's surface. Sometimes volcanoes erupt. This means liquid rock called lava explodes through the rip. Most of Venus is covered with hard lava.

Some areas of Venus are flat. Higher places rise above the flat land. These are called continents. A few mountains rise from the continents.

Scientists still have questions about Venus. Spacecraft will help them make new discoveries. These will help people learn more about both Venus and the solar system.

There are many large volcanoes on Venus.

Hands On: The Greenhouse Effect

The greenhouse effect makes Venus very hot. You can make a terrarium to grow small plants. This will show you how the greenhouse effect works.

What You Need
A glass or plastic jar with a lid
Potting soil
Ten plant or vegetable seeds
Water

What You Do
1. Take the lid off the jar.
2. Put two inches (five centimeters) of soil in the jar's bottom.
3. Sprinkle the seeds on the soil.
4. Cover the seeds with one inch (two and one-half centimeters) of soil.
5. Sprinkle a little water over the soil.
6. Place the lid back on the jar.
7. Place your jar in a sunny place.
8. Sprinkle a little water on the soil once a week.

The jar lid works like Venus' atmosphere. It will trap the Sun's heat. The inside of your terrarium will become very warm. This heat will help plants grow.

Words to Know

astronomer (uh-STRON-uh-mur)—person who studies stars, planets, and space

atmosphere(AT-muhss-fihr)—the mix of gases that surrounds some planets

greenhouse effect (GREEN-houss uh-FEKT)—the trapping of heat by a thick atmosphere

crater (KRAY-tur)—a hole in the ground made by a meteorite

outer space (OU-tur SPAYSS)—space outside a planet's atmosphere

phases (FAZ-ess)—the different parts of a moon or planet lit up by the Sun

rotation (roh-TAY-shuhn)—to spin around

spacecraft (SPAYSS-kraft)—craft built to travel in space

Read More

Barrett, N.S. *Planets*. New York: Franklin Watts, 1985.

Branley, Franklyn. *The Planets in Our Solar System*. New York: Harper & Row, 1981.

Simon, Seymour. *Venus*. New York: Morrow Junior Books, 1992.

Useful Addresses

NASA Headquarters
300 E Street SW
Washington, DC 20546

National Air & Space Museum
Smithsonian Institution
Washington, DC 20560

Internet Sites

Kids Web—Astronomy and Space
http://www.npac.syr.edu/textbook/kidsweb/astronomy.html

The NASA Homepage
http://www.nasa.gov/NASA_homepage.html

StarChild: A Learning Center for Young Astronomers
http://heasarc.gsfc.nasa.gov/docs/StarChild/StarChild.html

Index